Bibliographic information published by the German National Library:

The German National Library lists this publication in the National Bibliography; detailed bibliographic data are available on the Internet at http://dnb.dnb.de .

Imprint:

Print and binding: Books on Demand GmbH, Norderstedt Germany
ISBN: 9783668733633

This book at GRIN:

https://www.grin.com/document/427202

Andrew Choi

Improving the Daily Conditions of Life through Acoustics

GRIN Verlag

Improving the Daily Conditions of Life through Acoustics

Word Count: 4765

Improving the Daily Conditions of Life through Acoustics

Abstract:

Acoustics plays an integral role in human behavior and health. Not only are there useful applications of acoustics, such as using findings from research to predict Alzheimer's Disease or to induce different emotions such as happiness or sadness, but acoustics can also be used to potentially solve one of the most serious problems in the environment: noise pollution. Noise pollution is quickly becoming one of the biggest issues of the 21st century, and this has led to an alarming increase in the population affected a tinnitus, a medical condition that is related to hearing loss. In fact, "it has been estimated that 94 percent of the people in the United States suffer from some degree of tinnitus."[1] Furthermore, noise pollution greatly affects worker productivity and their health. In order to solve these problems and better society, many different steps have been taken to try and reduce the amount of noise pollution. Some include creating more productive office designs in order to counteract this problem or allowing ANC (active noise control) devices to be more accessible to the public. One aspect of acoustics that this study focuses on is psychoacoustics and understanding how acoustical properties of multiple environments affect behavior. This study will attempt to discover the potential correlation between noise pollution and human behavior. Acoustics and noise pollution are certainly intertwined, and if a connection exists between acoustics and human behavior, then it can be conjectured that a possible correlation can be created between noise pollution and human behavior. What this study intends to focus on is what factor specifically manipulates the correlation between acoustics and human behavior, and in order to explore this issue, a thorough meta-analysis will be conducted.

[1] Blanchfield, D. (2011). Noise pollution. *Gale: Science in Context.* Retrieved from http://link.galegroup.com/apps/doc/CV2644150947/SCIC?u=j043905010&xid=138ac323

Improving the Daily Conditions of Life through Acoustics

1.1 - Introduction:

Acoustics is defined as the physics of sound. Sound is observed as a mechanical vibration that disturbs the equilibrium of an elastic material medium, such as air or water.[2] Sound is something that surrounds people at all times, and it is one of the most dominating forces in the environment. For this reason, acoustics exists in order to explain how sound works and propagates itself through the environment. In this specific paper, I plan to delve into psychoacoustics, which is known as the branch of acoustics dealing with how sound psychologically affects the human brain and behavior.

It has already been proven in the past that sound--and acoustic engineering as a whole--has many different applications. Anna Barney, Professor of Biomedical Acoustic Engineering at the University of Southampton, discovered that analyzing speech patterns may be an indication of a form of dementia, such as Alzheimer's.[3] Furthermore, a study by Ana Tajadura-Jiménez, Pontus Larsson, et al. revealed that the room size of a room changes the emotional responses of people.[4] Both of these studies have revealed two important aspects of human nature and behavior in terms of acoustics. Psychoacoustics plays such a key role in daily life that various factors affecting people's behaviors can often be reduced to the sound in the environment.

1.2 - Noise Pollution:

One major problem that modern society faces today is the problem of noise pollution. The Environmental Encyclopedia, from Gale Databases, states that 94% of Americans have some degree of tinnitus, a hearing loss disorder.[5] Most of this tinnitus is caused by the excessive amounts of noise pollution in the environment. Some of these causes include--but are not limited

---

[2] Berg, R. (2017). Sound. *Britannica School.* Retrieved from http://school.eb.com/levels/high/article/sound/109557

[3] Barney, A. (2017). Alzheimer's: managing the conversation. *Southampton.ac.uk.* Retrieved from https://www.southampton.ac.uk/news/2017/03/alzheimers.page

[4] Tajadura-Jiménez, A., Larsson, P., Väljamäe, A., Västfjäll, D., & Kleiner, M. (2010). When room size matters: Acoustic influences on emotional responses to sounds. *Emotion, 10(3),* 416-422. http://dx.doi.org/10.1037/a0018423

[5] Blanchfield, D. (2011). Noise pollution. *Gale: Science in Context.* Retrieved from http://link.galegroup.com/apps/doc/CV2644150947/SCIC?u=j043905010&xid=138ac323.

to--cars, airplanes, loud music, or construction work. The problem here, however, is that according to the American Hearing Research Foundation, there is no treatment or cure for most types of hearing loss.[6] Perhaps if this medical condition was reversible or curable, then noise pollution would become less of an issue.

An interview between Stacy Fitzgerald-Redd and Jeremy Luscombe revealed that excessive amounts of noise--even background noise--can reduce productivity by 66% and also induce stress, which includes high blood pressure and a spike in heart rate.[7] This is only one aspect of the impact that noise pollution can have on the public. It not only undermines worker productivity, leading to less income and economic benefits, but can also negatively impact employees' lives.

1.3 - Potential Solutions:

Of course, even though there is no silver bullet to noise pollution, many steps are being taken to reduce the amount of damage that noise pollution can do. One study, conducted by Ya-Lin Yu, Yu-Tin Chao, Lai-Chung Lee, Jia-Yush Yen, and Yun-Chiu Fan, created an innovative office design implementing acoustical theorems such as the double wall principle or a pentagon-shaped design that would decrease the amount of noise pollution by a significant amount.[8] These are important steps in the monumental task of solving noise pollution. Another significant step included the creation of a more accessible and portable active noise control (ANC) device.[9] ANC devices are meant to reduce low-frequency sounds. These low sounds are often ignored, but these

[6] Runge, Christina. (2012). Noise Induced Hearing Loss. *American Hearing Loss Foundation.* Retreived from http://american-hearing.org/disorders/noise-induced-hearing-loss/

[7] Fitzgerald-Redd, S. (Interviewer) & Luscombe, J. (Interviewee). (1999). Acoustics Affect Productivity and Well Being [Interview transcript]. Retrieved from http://information.insulationinstitute.org/blog/noise-affects-productivity-and-well-being

[8] Yu, Y., Chao, Y., Lee, L., Yen, J., & Fan, Y. (2015). A Novel Soundproof Ventilation Plant Design with High Performance and No Energy Consumption. *Mathematical Problems In Engineering, 2015, 1-10.* http://dx.doi.org/10.1155/2015/496748

[9] Wang, C., Gao, H., Yu, L., Yu, T., Yan, W., & Xue, Q. (2016). Portable Low-Frequency Noise Reduction Device for Both Small Open and Closed Spaces. *Shock And Vibration, 2016, 1-11.* http://dx.doi.org/10.1155/2016/6241935

noises have a large impact on the subconscious and mental health of the people in the environment. These sounds are quite enigmatic, and it is extremely hard to reduce these types of sounds in the environment. For this reason, although this ANC device is still in its experimental stage, it is still an amazing step towards stopping the amount of noise pollution in the world today.

However, one solution that has not yet been implemented or researched completely is the use of acoustic engineering to reduce noise pollution in neighborhoods or public spaces. Although steps have been taken to reduce noise production of machinery and technology, noise pollution still takes such a prominent role in the environment. Many groups of people, such as the elderly, suffer largely from hearing loss due to the acoustical problems in the built environment.[10] Although some of the causes may come from deterioration of the ear or the fact that they were exposed to general amounts of noise over a longer time period, many of these old individuals stated that one of the major problems was poor speech communication, due to reverberant spaces and background noise. Applications of acoustics into the built environment has begun to be explored and may be the next step in solving this environmental issue of noise pollution.

One aspect of acoustics I wish to focus on this study, however, is psychoacoustics and how acoustics affects the brain and human behavior. It is not unknown how sound affects humans; music has been known to create a vast variety of emotional responses.[11] However, the effect that the acoustical properties of an environment have on human behavior is still quite unknown. This is the knowledge gap that I would like to explore for this study, as it is a part of psychoacoustics that is yet to be explored. If a certain correlation between acoustics of a room and human behavior exists, then a conjecture can be made that noise pollution definitely affects hu-

[10] Cox, T., Davies, B., Longhurst, B., & Webb, C. (2017). Hearing loss in the Built Environment. *University of Salford, Manchester.* Retrieved from http://www.salford.ac.uk/computing-science-engineering/research/acoustics/psychoacoustics/hearing-loss-in-the-built-environment

[11] Taylor, I., & Paperte, F. (1958). Current Theory and Research in the Effects of Music on Human Behavior. *The Journal Of Aesthetics And Art Criticism, 17*(2), 251-258. http://dx.doi.org/10.2307/427527

man behavior, and it makes noise pollution a problem that not only then concerns the environment but also our society as well.

2.1 - Research Design:

This paper aims to answer whether acoustics affects human behavior through changing auditory perception. In order to do so, the best and most intuitive way to answer this question is through experimental and correlational research. By directly using human participants, auditory stimuli, and different acoustical environments, some sort of correlation--or disassociation--will be observed. However, an experiment requires extensive lab equipment that is not available to students, and therefore, I will be unable to conduct an experiment. Instead, non-experimental research in the form of meta-analysis will be used in order to discover whether there is a correlation between acoustics and human behavior. This method is very reliable in creating a solid conclusion based on the material in different sources. This method is also used commonly in psychology, which is directly related to psychoacoustics.

The first thing to realize is that different auditory stimuli such as a dog barking, classical music, or a rock song evoke different emotions and responses. The classical music may cause relaxation while rock songs will incites excitement and energy. A jazz song may encourage dancing, or a dog barking could cause feelings of happiness or distress, depending on the tone of the dog.

These different sounds also have inherently different tempi, which have been known to cause different psychological effects. In a study published in the International Journal of Sports Medicine, different music tempi change the efficiency of exercise; medium tempi have the greatest effect on intrinsic motivation and have been shown to yield the best results in exercise.[12] Not only are the notes important, but the timing and the rhythm are important--significant enough to affect the psychological and physical response of humans.

The intensity of a sound also plays into the role of auditory perception. Donald H. Mershon and L. Edward King's study asserts that "Both auditory intensity and reverberation" are different properties that "produce systematically varying judgments of perceived distance when" the same auditory stimuli is repeated "to the same observer."[13] Evidently, the loudness and perceived magnitude of the sound will change the different reactions that the different participants will have. It is important to realize, however, that too much noise can be extremely harmful. According to Figure 1, prolonged exposure to 90 dB or higher is dangerous, and it can cause serious damage to one's ears without any sort of hearing protection. 90 dB is comparable to traffic, and even higher volumes are comparable to the sounds of loud rock concerts, jet engines, or a gun muzzle blast, which are loud enough that exposure should be limited as much as possible. Due to its great magnitude and sound intensity, it creates an uncomfortable feeling for the listener, which is another example that supports the logic that magnitude does affect human behavior and psychological responses.

Figure 1: Approximate Decibel Levels in Real World Examples.[14]

Approximate Decibel Level	**Examples**
0 dB	the quietest sound one can hear.
30 dB	whisper, quiet library
60 dB	normal conversation, sewing machine, typewriter.

[12] Karageorghis, C., Jones, L., & Stuart, D. (2008). Psychological Effects of Music Tempi during Exercise. *International Journal Of Sports Medicine, 29(7)*, 613-619. http://dx.doi.org/10.1055/s-2007-989266

[13] Mershon, D., & King, L. (1975). Intensity and reverberation as factors in the auditory perception of egocentric distance. *Perception & Psychophysics*, *18*(6), 409-415. http://dx.doi.org/10.3758/bf03204113

[14] Runge, Christina. (2012). Noise Induced Hearing Loss. *American Hearing Loss Foundation*. Retreived from http://american-hearing.org/disorders/noise-induced-hearing-loss/

90 dB	lawnmower, shop tools, truck traffic; 8 hours per day is the maximum exposure (protects 90% of people)
100 dB	chainsaw, pneumatic drill, snowmobile; 2 hours per day is the maximum exposure without protection.
115 dB	sandblasting, loud rock concert, auto horn; 15 minutes per day is the maximum exposure without protection.
140 dB	gun muzzle blast, jet engine; noise causes pain and even brief exposure injures unprotected ears; maximum allowed noise with hearing protector

2.2 - Hypotheses and Assumptions:

First of all, it is important to address that I will be assuming that acoustics will directly affect human behavior. In order to find a correlation between acoustics and human behavior, it needs to be assumed that there is already a direct relationship between the two. Previously, in the literature review, it was stated that room size does indeed change emotional responses to sounds--whether that includes intensifying those emotions or providing calmer reactions. Although room size is a big factor in room acoustics, it is also important to address the reflections, diffusions, and absorptions that the sound waves in the room are experiencing. In order to explore the effect of acoustics on human behavior, it is important to first address that these acoustical qualities will in fact even change human behavior.

One study published in the Journal of the Acoustical Society of America explores the effect of reverberation on the auditory perception of participants.[15] This specific experiment tested the participants through different trials of auditory stimuli in different environments, and as a result, the researchers were able to deduce that some used reverberation time or the direct/reverberant energy ratio to determine their bearings and the room size. Although this source

[15] Sandvad, J. (1999). Auditory perception of reverberant surroundings. *The Journal Of The Acoustical Society Of America, 105(2),* 1193-1193. http://dx.doi.org/10.1121/1.425625

does not discuss the psychological implications of their findings, it is clear that reverberation makes a huge difference in the human participants' mind.

One keen example of this effect--concerning how reverberation affects the auditory perception of humans--is personified in a study that explores the effects of acoustics on the performance of an employee in a working environment. These employees were instructed to complete a questionnaire in both a private office and an open-plan office, and the results undoubtedly showed a revealing trend about open-plan offices; compared to private offices, employees in the open-plan office displayed increased distraction, reduced privacy, increased concentration difficulties, and increased use of coping strategies.[16] The reason for this extremely negative response to the open-office may be due to the amount of increased social interaction and noise levels, but it is clear that the poor acoustics also plays into this negative response in the open-office design. The noise levels were amplified by the increased number of reverberations in the office, which leads to the amount of decreased worker efficiency. Whereas the open-plan office decreased the amount of work completed, the private-office system was much more successful in achieving workplace goals. This is one example that brings to light the effect acoustics on the psychological behavior of humans.

Through our research, we can find that if acoustics can be used to adjust the emotional responses of people, we can reduce the amount of noise pollution in our environment, we can propel the advancement of society and encourage cooperation. Acoustics is already used and applied to business conference rooms and others, and applying this to daily life would benefit all people. It will make acoustics have another meaningful application in our environment. Not only will acoustics be able to improve human behavior in modern society, but it will also strengthen

[16] Kaarlela-Tuomaala, A., Helenius, R., Keskinen, E., & Hongisto, V. (2009). Effects of acoustic environment on work in private office rooms and open-plan offices - longitudinal study during relocation. *Ergonomics*, *52*(11), 1423-1444. http://dx.doi.org/10.1080/00140130903154579

the importance of reducing the amount of noise pollution. The amount of noise directly correlates to auditory perception and psychological behavior; as the noise levels of a room increases, the auditory perception diminishes and psychological behavior generally worsens; the trend is mostly negative as the intensity level of the noise increases.[17]

2.3 - Obstacles

It is important to realize that there are many limitations and potential errors in this study. Due to the use of non-experimental research, there are definitely limitations in the sense that there may not be any studies or information about certain aspects of my research. One obstacle that could impede my research is simply the lack of information in this field. This should not serve as a major problem to my research, but it is important to address it as one of the limitations of this study.

Another potential margin for error is the use of generalizations. In order to use meta-analysis, generalizations must be made on certain quantities across multiple sources in order to find a significant trend or correlation. Therefore, although the analysis may provide a big picture of the issue, it does not include the entire picture of the issue. It only contains a portion of all of the data in order to make a general consensus and conclusion to the data.

Furthermore, the use of correlational research adds a degree of uncertainty in my data; although it is a common method used in psychology and is often the most recommended method to pursue this kind of study, it can all crumble due to this statement: correlation is not causality. It might seem that there is a correlation between two separate variables, but in actuality, it may have been a coincidence. However, my research mainly focuses on the possibility of where there might be a variable that correlates between noise pollution and human behavior, and I believe that my research and correlational methods will be sufficient enough to answer this question.

[17] Azrin, N. (1958). Some Effects of Noise on Human Behavior. *Journal Of The Experimental Analysis Of Behavior, 1*(2), 183-200. http://dx.doi.org/10.1901/jeab.1958.1-183

3.1 - Initial Connections:

The first step in understanding how acoustics affects behavior requires the knowledge of how music or sounds affect human behavior. Through multiple sources, it is found that the effects of music are many; they can range from increased amounts of cortisone, testosterone, and oxytocin to the change in metabolism, respiration, or even patterns of brain waves.[18,19] Therefore, the effects of music in changes of the body are very prevalent. Along with the modified patterns of brain waves, these work together to create a significant change in human behavior. This is supported by the fact that musicians have a greater amount of gray matter in their brain compared to nonmusicians due to the acquisition or development motor and auditory skills.[20]

Now that there is a basic understanding of how exactly music affects human behavior, we need to next understand the effect that acoustics of an environment has on music. Obviously, concert hall acoustics has different properties from a studio room, and these acoustical properties must be analyzed to understand how they affect music. There exist many different acoustical properties, such as reflection or diffraction, but the acoustical property that I will focus on in this study will be reverberation, which is mainly dependent on the size of the given space in which the sound is propagated in and the materials. These will noticeably change how exactly a sound will propagate itself over the course of time. As seen in Figure 2 below, the different environments vary substantially in size and reverberation time, and this is what gives them that distinct sound in each environment. Music largely depends on the reverberation time, denoted by RT60, which refers to the amount of time it takes for a sound to decay 60 dB. This dictates the amount of time a sound seems to stay in the air. The materials also make a huge difference in affecting

[18] Taylor, I., & Paperte, F. (1958). Current Theory and Research in the Effects of Music on Human Behavior. *The Journal Of Aesthetics And Art Criticism, 17*(2), 251. http://dx.doi.org/10.2307/427527
[19] Lemonick, M. (2000). *Music on the Brain. TIME.com.* Retrieved from http://content.time.com/time/magazine/article/0,9171,46157,00.html
[20] Gaser, C., & Schlaug, G. (2003). Gray Matter Differences between Musicians and Nonmusicians. *Annals Of The New York Academy Of Sciences, 999*(1), 514-517. http://dx.doi.org/10.1196/annals.1284.062

the reverberation time and how exactly the sound interacts with the environment; acoustic foam is known to absorb noise, and this leads to a clearer sound compared to concrete which just reflects the noise back, creating more acoustic chaos in the environment. It can make beautiful notes linger in a concert hall, or it can release a symphony of cacophonies in an echoing gymnasium.

Figure 2: Common Acoustical Environments[21,22,23,24]

Environment	Reverberation Time (RT60)	Size (approximate; in thousands of cubic ft)	Materials
Concert Hall	2 or more secs.	300 - 900	Wood, concrete, steel
Classroom	Less than .6 secs.	1 - 2	Wood, concrete, plaster, glass, carpet
Gym	1 - 2 secs	50 - 200	Wood, concrete, plastic, glass, rubber
Office	.5 - 1 secs	1 - 2.5	Brick, Wood, glass, concrete, drywall

Now it is clear how exactly acoustics affects music; the reverberation time and mechanical properties of the materials influence the sound and music in the environment. Now comes the important part where we connect the dots: how does the acoustics affect human behavior? It is already clear that the prevalence of noise will create a more detrimental effect; this is seen in Figure 3 down below. Therefore, it is logical to say that if the acoustics is adjusted so that there less reverberation time (as this quantity largely controls the amount of lingering sound, manipulating

[21] *The Concert Hall.* (2018). *Aac.pref.aichi.jp.* Retrieved from http://www.aac.pref.aichi.jp/english/gekijyo/guide/c_holeg.html
[22] Dyer, R. (2005). *The Concert Hall. Meyerson Symphony Center.* Retrieved from http://meyerson.dallasculture.org/about/the-concert-hall/
[23] *American National Standard on Classroom Acoustics.* (2002). *Asha.org.* Retrieved from https://www.asha.org/public/hearing/American-National-Standard-on-Classroom-Acoustics/
[24] *Acoustic Information for Gymnasiums.* (2003). *Acoustics.com.* Retrieved 3 April 2018, from http://www.acoustics.com/gym.asp

the amount of total sound), it will likely improve human behavior. This can mean improvements in social conduct or increased academic performance.

Figure 3: Potential Negative Effects of Noise Pollution[25]

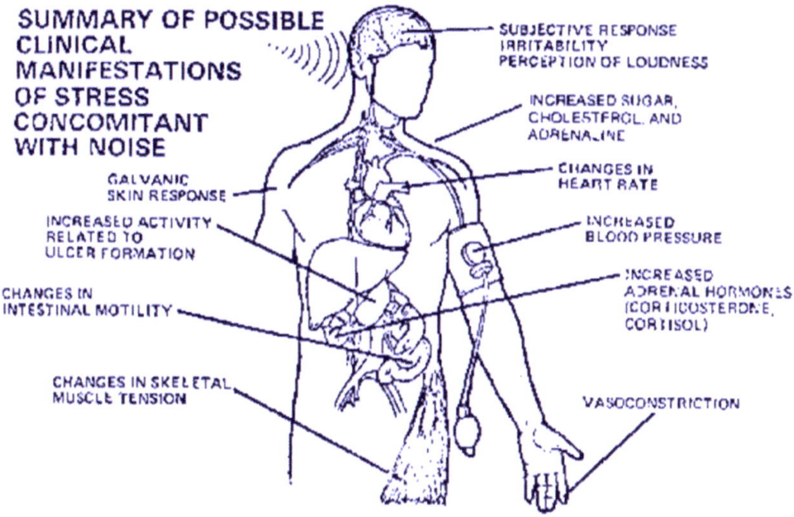

3.2 - The Classroom:

A compilation of different studies about the classroom environment and its acoustics proves just that. 5 separate studies, although they focus on different aspects of the classroom environment, prove the underlying argument that acoustics are central to a student's learning and behavior. The literature review recognizes that the classroom is contingent on many factors, of which is the architecture and design of the classroom.[26] This is a good introduction to the data, and the article published by Woolner, Hall, Higgins, McCaughey, and Wall all proves a point that

[25] *Noise Effects Handbook.* (1981). *NPC Library.* Retrieved from https://www.nonoise.org/library/handbook/handbook.htm

[26] Higgins, S., Hall, E., Wall, K., Woolner, P., & McCaughey, C. (2005). The impact of school environments: A literature review. *London: Design Council.*

this amount of noise may lead to irritation, loss of concentration, and a lack of learning.[27] The next two sources, however, use numerical figures to prove their point of how important acoustics are to the educational environment. It is seen that in the study authored by Crandell and Smaldino, a trend was seen that the test results decrease as the distance from the noise source (the teacher) decreases.[28] In the fourth study, a trend is seen that as a student gets older, the effect that the acoustics has on the student decreases and the audibility increases.[29] This is extremely important, as the students get older, test results will have a smaller differential based on the distance from the teacher, and the effect that acoustics has on the students' test results decreases. Therefore, it is logical to say that acoustics in the classroom may not play that big of a role in higher education, but due to the results in Crandell and Smaldino's study, acoustics in the classroom environment are important in the primary school environment. This is especially corroborated by the conclusions of Dockrell and Shield's study about acoustical barriers in the classroom environment; not only does the acoustics of the classroom lead to decreased academic performance, but the prevalence of noise also led to negative trends in subjects reading and spelling.[30]

An important conclusion from this meta-analysis is that acoustics play a less significant role in learning and understanding as one increases with age. Although this is significant, there exists one final conclusion which directly answers the question that I am answering: as audibility decreases, the academic performance also suffers. This is especially since in Crandell and Smaldino's study in which the test results worsen as the distance from the student to the teacher in-

[27] Woolner, P., Hall, E., Higgins, S., McCaughey, C., & Wall, K. (2007). A sound foundation? What we know about the impact of environments on learning and the implications for Building Schools for the Future. *Oxford Review of Education*, *33*(1), 47-70.

[28] Crandell, C. C., & Smaldino, J. J. (2000). Classroom acoustics for children with normal hearing and with hearing impairment. *Language, speech, and hearing services in schools*, *31*(4), 362-370.

[29] Nelson, P. B., & Soli, S. (2000). Acoustical barriers to learning: Children at risk in every classroom. *Language, speech, and hearing services in schools*, *31*(4), 356-361.

[30] Dockrell, J. E., & Shield, B. M. (2006). Acoustical barriers in classrooms: The impact of noise on performance in the classroom. *British Educational Research Journal*, *32*(3), 509-525.

creases.[31] This directly shows how the acoustical properties of a room affect the performance and behavior of a person; the reverberation time directly affects the audibility of a person in a given environment, and this shows that reverberation time has an inverse relationship with behavior, as the greater reverberation time creates less speech intelligibility. This is seen especially in how a greater reverberation time encourages greater amounts of sound, which result in lower test scores and academic proficiency.

3.3 - The Gym:

This correlation found in the classroom is also inherent in the gymnasium environment. The first concept that must be understood is the acoustical environments of the gymnasiums found throughout the entire nation. According to a study conducted and published in 2008, only about 5% of the gymnasiums actually met the noise requirements of the recommended noise level.[32] This means that the majority of gymnasiums all across the nation have a reverberation time of over the recommended noise levels of 1-2 seconds--some even have levels of 4 or more seconds of reverberation time--and this discovery is very alarming. A reverberation time of 1.5 seconds is optimal, as it maximizes the amount of energy in an environment without compromising too much speech intelligibility (which is especially important during games or school assemblies in a gym), but since gyms have reverberation times over 2 seconds, the level of speech intelligibility and audibility is significantly crippled.[33]

Due to the fact that the audibility is significantly impaired, this leads to two different scenarios: the overall atmosphere of the gym is very energetic or interpersonal relationships deteriorate due to the amount of frustration birthed from the inability of communication. Similarly, in

[31] Crandell, C. C., & Smaldino, J. J. (2000). Classroom acoustics for children with normal hearing and with hearing impairment. *Language, speech, and hearing services in schools*, *31*(4), 362-370.

[32] Ryan, S., & Mendel, L. (2010). Acoustics in physical education settings: the learning roadblock. *Physical Education & Sport Pedagogy*, *15*(1), 71-83. http://dx.doi.org/10.1080/17408980902729370

[33] *The Acoustical Challenges of Gyms with Auditorium and Classroom Uses.* (2013). *Acoustical Engineering: Noise Control and Vibration Abatement.* Retrieved from http://www.acoustical-consultants.com/acoustical-challenges-of-mixed-use-school-gym-auditoriums-classrooms/

the classroom environment, the amount of performance was affected by the amount of reverberation time or the audibility between the student and teacher. It especially proves true in this study in which the reduction of reverberation time led to the improvement in overall behavior and demeanor.[34] Furthermore, the increased noise exposure due to the unusually high reverberation times also leads to a significant amount of hearing loss.[35] Already it is clear that if there is a high amount of noise pollution and hearing loss, it leads to detrimental health effects, such as a higher risk for cardiovascular disease (see Figure 3 above). Overall, it is clear that through the amount of reverberation time in a gymnasium environment, the overall behavior and conduct is negatively impacted.

3.4 - The Office:

The last acoustical environment that will be explored in this study is the office environment, or, put more broadly, the working environment. Earlier in the paper, the difference between the open-office and the private office and the relative effectiveness of the two was discussed, and this initial discovery will be further discussed here. One of the main reasons why the private office prevailed in terms of effectiveness over the open office was due to the absence of a high reverberation time.[36] Perhaps it was also due to a greater amount of privacy or lack of distractions, but all of this stems from the small reverberation time. The small amounts of noise accumulate in the office environment due to the reverberation time and these small noises contribute to "quasi-noise pollution" in the environment, which fosters the seeds of chaos and unproductivity.

This rings true in other studies about the office environment. One model shows that the amount of speech intelligibility--which reverberation time directly affects--directly correlates

[34] Bistrup, M. (2003). Prevention of adverse effects of noise on children. *Noise & Health*, *5*(19), 59-64.
[35] Maffei, L., Iannace, G., Masullo, M., & Nataletti, P. (2009). Noise exposure in school gymnasia and swimming pools. *Noise Control Engineering Journal*, *57*(6), 603. http://dx.doi.org/10.3397/1.3155385
[36] Kaarlela-Tuomaala, A., Helenius, R., Keskinen, E., & Hongisto, V. (2009). Effects of acoustic environment on work in private office rooms and open-plan offices - longitudinal study during relocation. *Ergonomics*, *52*(11), 1423-1444. http://dx.doi.org/10.1080/00140130903154579

with the amount of productivity.[37] A study focusing on a small office model shows that the more absorbent rooms had a higher sound localization score compared to the more reverberant rooms; the absorbent rooms provided more audibility than the reverberant rooms.[38] Both of these show the importance and relevance of reducing the amount of reverberation time in order to improve the amount of worker productivity.

One source from my meta-analysis seems to state the contrary: minimizing speech intelligibility increases the amount of worker productivity.[39] Although this encourages increasing the amount of reverberation time, it is actually very clear that this is done in order to reduce the number of side conversations and distractions in the open-office environment. The underlying correlation is still clear: manipulating the amount of reverberation time in order to maximize the amount of productivity.

4.1 - Discussion and Justifications:

This study's entire purpose was to examine whether the acoustical properties of the environment affects auditory perception and human behavior in any way. Through my meta-analysis, the underlying factor in determining the relationship between acoustics and human behavior is very clear; this factor is reverberation time. In the classroom, it was clear that a smaller reverberation time is optimal. In the gym, a higher reverberation time is acceptable, but too high leads to negative impacts on health and psychological behavior. In the office, reverberation times can be high or small, but preferably small in order to foster speech intelligibility but also minimal

[37] Hongisto, V. (2005). A model predicting the effect of speech of varying intelligibility on work performance. *Indoor Air, 15*(6), 458-468. http://dx.doi.org/10.1111/j.1600-0668.2005.00391.x

[38] Giguère, C., & Abel, S. (1993). Sound localization: Effects of reverberation time, speaker array, stimulus frequency, and stimulus rise/decay. *The Journal Of The Acoustical Society Of America, 94*(2), 769-776. http://dx.doi.org/10.1121/1.408206

[39] Jahncke, H., Hongisto, V., & Virjonen, P. (2013). Cognitive performance during irrelevant speech: Effects of speech intelligibility and office-task characteristics. *Applied Acoustics, 74*(3), 307-316. http://dx.doi.org/10.1016/j.apacoust.2012.08.007

noise pollution. Whether the reverberation time affects the behavior positively or negatively, it is clear that this is one of the major factors in determining the relationship between acoustics and human behavior.

Although I had first considered experimental and correlational research, as they are one of the best ways to approach this question and problem, I eventually realized that I would need extensive lab equipment in order to truly analyze the emotional responses of the participants. That is why I opted for meta-analysis instead, a method that is commonly used in psychology. Although predicting the patterns of the brain has been a field that has been studied over centuries, and even with enhanced technology or improved scientific approaches, the human brain is an extremely advanced supercomputer that has and could never be completely mapped for at least the next century or so, psychology has surely improved, and it can help to generalize the patterns of human behavior. In such an obscure field as psychoacoustics, it is probable that an experiment may be more effective, but it is better to opt for a more non-experimental approach due to the lack of materials to conduct the experiment effectively. However, if I were to pursue an experimental research study concerning acoustic perception, I would probably create an experiment where I test 5 auditory stimuli in 3 different acoustical environments and have the participants be exposed to them and record their response. This would focus on observing how their emotional responses changed according to the different acoustics. Perhaps a complicated 3D model to simulate acoustic environments would be also used in conjunction with this research method. Also, for the experiment, I would opt for a more free-response survey. Although making the participants scale their reactions on a numerical scale from 1 to 5 or 1 to 10 would make data analysis simple, this method is majorly flawed and has numerous limitations. Not only does it limit the amount of freedom that the participants could express about their reaction to the sound, but it is also subliminally influencing them to think or react in a certain way. This would definitely skew results and seem much more inflated than they actually were. So, opting for a more free-response questionnaire in which the participants could just state how they felt with minimal external influence would be much more effective.

4.2 - Applications:

This study can be applied to real-world scenarios today. First and foremost, many highways and roads today use walls in an attempt to reduce the amount of sound from propagating into neighborhoods, parks, and other public spaces. This is a good step in the right direction, but it still lacks a lot of sound absorption, and it is often ineffective. One group of researchers (Celia A., Carlos, L., et. al), in an attempt to improve the amount of sound absorption, have used co-combustion bottom ash to create a more effective sound barrier on highways.[40] Many efforts, especially this one, have been better than their commercial counterparts, and this is a revolutionary step in the process of reducing noise pollution. Using the knowledge gained from the research, we can find that reducing the amount of noise pollution that traffic creates in our daily lives will lead to greater worker productivity, improved mental health, and a higher quality and standard of living. Imagine the possibilities if the human race had discovered what my research has found: perhaps this greater sense of productivity would have led to the creation of a cure for cancer or the creation of smarter smartphone technologies.

It is also important to recognize applications of the different results from the acoustical environments. Based on the previous research that I have conducted, I can firmly conclude that classrooms should be designed in such a way that the reverberation time should be limited and that the distance between the student and teacher should be reduced. This can mean using different types of materials or encouraging the use of acoustic panels for proper acoustical treatment of

[40] Arenas, C., Leiva, C., Vilches, L., & Cifuentes, H. (2013). Use of co-combustion bottom ash to design an acoustic absorbing material for highway noise barriers. *Waste Management, 33(11)*, 2316-2321. http://dx.doi.org/10.1016/j.wasman.2013.07.008

these environments. For the gym, the regulations put in place--for the acoustical requirements--need to be more heavily enforced. A reverberation time of 1.5 seconds is the optimal quantity, and this should be put into every single gymnasium around the world. Lastly, the office is potentially one of the most important environments that needs to be improved. A private office needs to be pursued, and although it may cost more money to implement, it will provide more income and profit in the long run. The applications of this study prove to be many, and they can be implemented into our society without delay.

4.3 - Conclusions:

From this study, it is clear that acoustics does indeed affect human behavior. Whether it is in a classroom environment or a concert hall, the acoustics will always play a role in amplifying or harming human behavior. This happens through the fact that acoustics affects the auditory perception of humans, leading to the manipulation of audibility, the ability to discern sounds from noises and understanding the speech of the other person. Acoustics is able to do that through the manipulation of the reverberation time. More reverberation time means more obscurity in the exchange of speech between people. Sometimes, that is a good thing; it allows for sound to blend in a concert hall setting. However, it can also lead to increased irritability and loss of concentration, as seen in the meta-analysis of different acoustical environments. Overall, acoustics is a part of the built environment that has never been ignored in the past, and due to this new discovery, it is more important than ever to secure the best acoustics for an environment.

References:

Acoustic Information for Gymnasiums. (2003). *Acoustics.com.* Retrieved from http://www.acoustics.com/gym.asp

American National Standard on Classroom Acoustics. (2002). *Asha.org.* Retrieved from https://www.asha.org/public/hearing/American-National-Standard-on-Classroom-Acoustics

Arenas, C., Leiva, C., Vilches, L., & Cifuentes, H. (2013). Use of co-combustion bottom ash to design an acoustic absorbing material for highway noise barriers. *Waste Management, 33(11),* 2316-2321. http://dx.doi.org/10.1016/j.wasman.2013.07.008

Azrin, N. (1958). Some Effects of Noise on Human Behavior. *Journal Of The Experimental Analysis Of Behavior, 1*(2), 183-200. http://dx.doi.org/10.1901/jeab.1958.1-183

Barney, A. (2017). Alzheimer's: managing the conversation. *Southampton.ac.uk.* Retrieved from https://www.southampton.ac.uk/news/2017/03/alzheimers.page

Berg, R. (2017). Sound. *Britannica School.* Retrieved from http://school.eb.com/levels/high/article/sound/109557

Bistrup, M. (2003). Prevention of adverse effects of noise on children. *Noise & Health, 5*(19), 59-64.

Blanchfield, D. (2011). Noise pollution. *Gale: Science in Context.* Retrieved from http://link.galegroup.com/apps/doc/CV2644150947/SCIC?u=j043905010&xid=138ac323

Cox, T., Davies, B., Longhurst, B., & Webb, C. (2017). Hearing loss in the Built Environment. *University of Salford, Manchester.* Retrieved from http://www.salford.ac.uk/computing-science-engineering/research/acoustics/psychoacoustics/hearing-loss-in-the-built-environment

Crandell, C. C., & Smaldino, J. J. (2000). Classroom acoustics for children with normal hearing and with hearing impairment. *Language, speech, and hearing services in schools, 31(4)*, 362-370.

Dockrell, J. E., & Shield, B. M. (2006). Acoustical barriers in classrooms: The impact of noise on performance in the classroom. *British Educational Research Journal, 32(3)*, 509-525.

Dyer, R. (2005). The Concert Hall. *Meyerson Symphony Center*. Retrieved from http://meyerson.dallasculture.org/about/the-concert-hall/

Elliott, L., Hammer, M., & Evan, K. (1987). Perception of gated, highly familiar spoken mono-syllabic nouns by children, teenagers, and older adults. *Perception & Psychophysics, 42(2),* 150-157. http://dx.doi.org/10.3758/bf03210503

Fitzgerald-Redd, S. (Interviewer) & Luscombe, J. (Interviewee). (1999). Acoustics Affect Productivity and Well Being [Interview transcript]. Retrieved from http://information.insulationinstitute.org/blog/noise-affects-productivity-and-well-being

Gaser, C., & Schlaug, G. (2003). Gray Matter Differences between Musicians and Nonmusicians. *Annals Of The New York Academy Of Sciences*, *999*(1), 514-517. http://dx.doi.org/10.1196/annals.1284.062

Giguère, C., & Abel, S. (1993). Sound localization: Effects of reverberation time, speaker array, stimulus frequency, and stimulus rise/decay. *The Journal Of The Acoustical Society Of America*, *94*(2), 769-776. http://dx.doi.org/10.1121/1.408206

Higgins, S., Hall, E., Wall, K., Woolner, P., & McCaughey, C. (2005). The impact of school environments: A literature review. *London: Design Council.*

Hongisto, V. (2005). A model predicting the effect of speech of varying intelligibility on work performance. *Indoor Air*, *15*(6), 458-468. http://dx.doi.org/10.1111/j.1600-0668.2005.00391.x

Jahncke, H., Hongisto, V., & Virjonen, P. (2013). Cognitive performance during irrelevant speech: Effects of speech intelligibility and office-task characteristics. *Applied Acoustics*, *74*(3), 307-316. http://dx.doi.org/10.1016/j.apacoust.2012.08.007

Kaarlela-Tuomaala, A., Helenius, R., Keskinen, E., & Hongisto, V. (2009). Effects of acoustic environment on work in private office rooms and open-plan offices - longitudinal study during relocation. *Ergonomics*, *52*(11), 1423-1444. http://dx.doi.org/10.1080/00140130903154579

Karageorghis, C., Jones, L., & Stuart, D. (2008). Psychological Effects of Music Tempi during Exercise. *International Journal Of Sports Medicine, 29(7)*, 613-619. http://dx.doi.org/10.1055/s-2007-989266

Lemonick, M. (2000). *Music on the Brain. TIME.com*. Retrieved 3 April 2018, from http://content.time.com/time/magazine/article/0,9171,46157,00.html

Maffei, L., Iannace, G., Masullo, M., & Nataletti, P. (2009). Noise exposure in school gymnasia and swimming pools. *Noise Control Engineering Journal*, *57*(6), 603. http://dx.doi.org/10.3397/1.3155385

Mershon, D., & King, L. (1975). Intensity and reverberation as factors in the auditory perception of egocentric distance. *Perception & Psychophysics*, *18*(6), 409-415. http://dx.doi.org/10.3758/bf03204113

Nelson, P. B., & Soli, S. (2000). Acoustical barriers to learning: Children at risk in every classroom. *Language, speech, and hearing services in schools, 31(4)*, 356-361.

Noise Effects Handbook. (1981). *NPC Library*. Retrieved 3 April 2018, from https://www.nonoise.org/library/handbook/handbook.htm

Runge, Christina. (2012). Noise Induced Hearing Loss. *American Hearing Loss Foundation*. Retreived from http://american-hearing.org/disorders/noise-induced-hearing-loss/

Ryan, S., & Mendel, L. (2010). Acoustics in physical education settings: the learning roadblock. *Physical Education & Sport Pedagogy*, *15*(1), 71-83. http://dx.doi.org/10.1080/17408980902729370

Sandvad, J. (1999). Auditory perception of reverberant surroundings. *The Journal Of The Acoustical Society Of America, 105(2)*, 1193-1193. http://dx.doi.org/10.1121/1.425625

Tajadura-Jiménez, A., Larsson, P., Väljamäe, A., Västfjäll, D., & Kleiner, M. (2010). When room size matters: Acoustic influences on emotional responses to sounds. *Emotion, 10(3)*, 416-422. http://dx.doi.org/10.1037/a0018423

Taylor, I., & Paperte, F. (1958). Current Theory and Research in the Effects of Music on Human Behavior. *The Journal Of Aesthetics And Art Criticism, 17*(2), 251. http://dx.doi.org/10.2307/427527

The Acoustical Challenges of Gyms with Auditorium and Classroom Uses. (2013). *Acoustical Engineering: Noise Control and Vibration Abatement*. Retrieved from http://www.acoustical-consultants.com/acoustical-challenges-of-mixed-use-school-gym-auditoriums-classrooms/

The Concert Hall. (2018). *Aac.pref.aichi.jp*. Retrieved 3 April 2018, from http://www.aac.pref.aichi.jp/english/gekijyo/guide/c_holeg.html

Wang, C., Gao, H., Yu, L., Yu, T., Yan, W., & Xue, Q. (2016). Portable Low-Frequency Noise Reduction Device for Both Small Open and Closed Spaces. *Shock And Vibration, 2016*, 1-11. http://dx.doi.org/10.1155/2016/6241935

Woolner, P., Hall, E., Higgins, S., McCaughey, C., & Wall, K. (2007). A sound foundation? What we know about the impact of environments on learning and the implications for Building Schools for the Future. *Oxford Review of Education, 33(1)*, 47-70.

Yu, Y., Chao, Y., Lee, L., Yen, J., & Fan, Y. (2015). A Novel Soundproof Ventilation Plant Design with High Performance and No Energy Consumption. *Mathematical Problems In Engineering, 2015*, 1-10. http://dx.doi.org/10.1155/2015/496748

Bibliografische Information der Deutschen Nationalbibliothek:

Die Deutsche Bibliothek verzeichnet diese Publikation in der Deutschen Nationalbibliografie; detaillierte bibliografische Daten sind im Internet über http://dnb.d-nb.de/ abrufbar.

Impressum:

Copyright © 2017 GRIN Verlag
Druck und Bindung: Books on Demand GmbH, Norderstedt Germany
ISBN: 9783668714939

Dieses Buch bei GRIN:

https://www.grin.com/document/420433

Stefan Nothdurft

Messen elektrischer Größen und PC-Messdatenerfassung mit LabVIEW

Laborbericht

GRIN Verlag

Stefan Nothdurft

Laborbericht

Messen elektrischer Größen und PC-Messdatenerfassung mit LabVIEW

Studiengang: Elektro- und Informationstechnik – Bachelor of Engineering (B. Eng.)

Modul: MSR41 – MST02

Eingereicht am: 16.10.2017

Inhaltsverzeichnis

Abbildungsverzeichnis

Tabellenverzeichnis

Abkürzungsverzeichnis

BNC	Bayonet Neill Concelman
COM	Common
DA	Data acquisition
GUI	graphical user interface
Vi	Virtuelle Instrumente

1 Einleitung

1.1 Ziel und Aufbau der Arbeit

Während einem zweitägigen Laborunterricht wurden diverse Messungen durchgeführt. In diesem Laborbericht werden die dort durchgeführten Messaufgaben im Detail beschrieben und dokumentiert. Zuerst wird die vorgegebene Aufgabenstellung definiert. Im Anschluss daran, ist die Durchführung der Messaufgaben der Reihenfolge nach beschrieben.

1.2 Aufgabenstellung im Überblick

Messen mit dem Digital-Multimeter

- Leerlaufspannung eines Solarzellenmoduls
- Kurzschlussstrom eines Solarzellenmoduls
- Spannungsmessung im Arbeitspunkt
- Leistungsabgabe eines Solarzellenmoduls bei Belastung
- Widerstand einer Motorwicklung

Messen mit dem Digital-Speicher-Oszilloskop

- Messung einer sinusförmigen Spannung
- Messung einer rechteckförmigen Spannung
- Spannungsmessung eines Solarzellenmoduls

Programmierung mit LabVIEW

- Generieren und Anzeigen von Zufallszahlen
- Inkrementieren eines Zählers mit einem Button
- Stoppuhr
- Würfel mit Statistik
- Umrechnung von Dezimalzahlen in das neuner Zahlensystem
    - Optional: Ohne Vi „Quotient & Remainder“

Messen und Auswerten mit LabVIEW

- Kalibrierung eines Wegsensors inkl. Parameter
- Kalibrierung eines Temperatursensors (Unterschiede zum Wegsensor)
- Auswertung einer Lichtschranke

2 Messen mit dem digitalen Multimeter

2.1 Digitales Multimeter ELV JT 138

Das digitale Multimeter wird verwendet um elektrische Größen zu erfassen. Es verfügt im Vergleich zum analogen Multimeter eine digitale Messwertanzeige und erleichtert dadurch das Ablesen des gemessenen Wertes. Im Labor stehen digitale Multimeter vom Typ JT 138 der Firma ELV zur Verfügung.

2.1.1 Anschluss

Der Anschluss der Messleitungen erfolgt über die Anschlussbuchsen auf der Vorderseite. Da Spannung und Strom unterschiedlich gemessen werden, sind mehrere Anschlussvarianten möglich. Die Buchse „V/Ohm" ist für die Spannungs- und Widerstandsmessung. Für die Strommessung stehen die beiden Messbereiche 20A und 200mA zur Verfügung. Die Buchse mit der Beschriftung „COM" ist der gemeinsame Masseanschluss.

Um den Überblick zu behalten werden Versorgungsspannung mit roten, Signale mit gelben und die Masse mit schwarzen Leitungen verbunden.

Keinesfalls darf eine Spannungsmessung über die Strommesseingänge durchgeführt werden, dadurch können Bauteile in der Schaltung oder das Messgerät selbst beschädigt werden.

Abbildung wurde aus datenschutzrechtlichen Gründen von der Redaktion entfernt.

Abbildung 1: Digitales Multimeter JT138

2.1.2 Digit und Auflösung

Ein Digit entspricht der niederwertigsten Stellenanzeige auf dem Display. Die Auflösung ist gleich einem Digit.

Beispiel: Messbereich 0,00V - 20,00V => 1 Digit = 0,01V, Auflösung = 0,01V

2.1.3 Messbereiche und Messgenauigkeit

Aus dem Datenblatt kann die Messgenauigkeit entnommen werden. Diese Angabe bezieht sich auf den angezeigten Messwert.

Der Messbereich des Multimeters wird vor der ersten Messung auf die maximal zu erwartende Spannung eingestellt. Anschließend wird der Messbereich solange angepasst, bis der Messwert im oberen Drittel des Messbereichs liegt. In diesem Bereich sind die Werte prozentual genauer.

Beispiel: Der Messbereich ist 20VDC, die Messabweichung laut Datenblatt +-(0,05% + 3d), Auflösung 1mV

Absolute Messabweichung bei 20V = +-0,10V , relative Messabweichung = +-0,05%

Absolute Messabweichung bei 1V = +-0,008V , relative Messabweichung = +-0,8%

2.2 Messung der Leerlaufspannung

Es wird die Leerlaufspannung eines Solarzellenmoduls gemessen werden.

Da die Helligkeit im Messlabor nicht ausreicht um die maximale Spannung zu erzeugen, wird eine handelsübliche Glühlampe auf das Solarzellenmodul ausgerichtet. Bei der Leerlaufspannungsmessung wird das Multimeter parallel an das Solarzellenmodul angeschlossen. Es darf kein weiterer Verbraucher angeschlossen sein. Das Multimeter besitzt im Spannungsmessbereich einen sehr hohen Innenwiderstand und ist in diesem Fall nicht als Verbraucher zu werten.

Es wurde der Messbereich 20V und die Gleichspannungsmessung eingestellt.

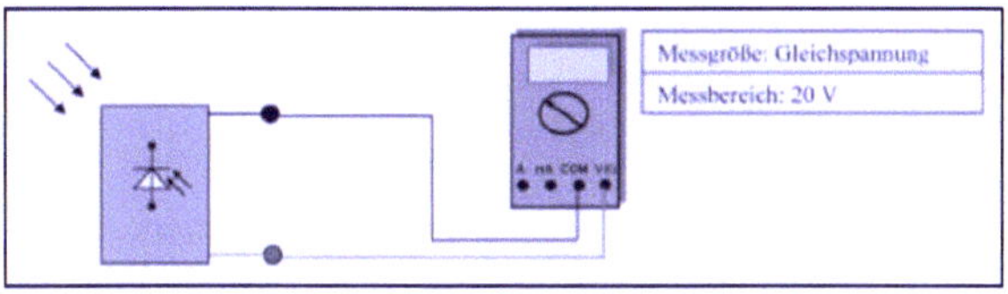

Abbildung 2: Messaufbau einer Leerlaufspannungsmessung

Messergebnis Berechnung

$U_{Leerlauf} = U_{Messung}$ +-(0,5% + 3d) = 2,42V +-(0,0121V+0,03V) = 2,42V +- 0,04V

Die Solarzelle liefert im Leerlauf eine Spannung von 2,42V +- 0,04V.

2.3 Messung des Kurzschlussstroms

Es wird der Kurzschlussstrom eines Solarzellenmoduls gemessen werden.

Bei der Kurzschlussstrommessung bleibt das Multimeter ebenso wie bei der Leerlaufspannungsmessung angeschlossen, jedoch werden die Buchsen „A“ und „COM“ verwendet. Das Solarzellenmodul liefert dann den maximalen Strom.

Der Messbereich ist auf 20A Gleichstrommessung eingestellt.

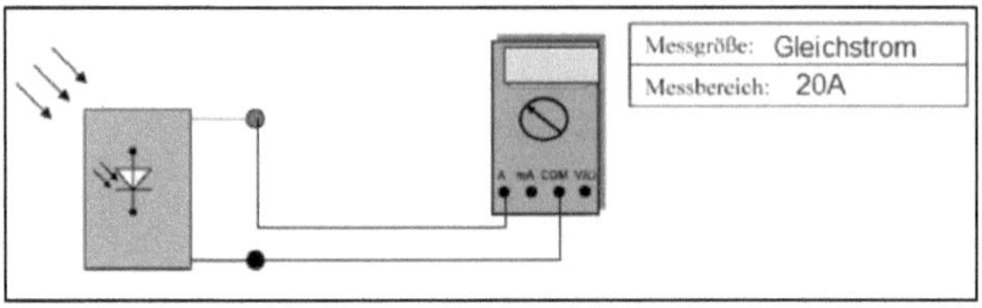

Abbildung 3: Messaufbau einer Kurzschlussstrommessung

Messergebnis Berechnung

$I_{Kurzschluss} = I_{Messung}$ +-(2,0% + 5d) = 360mA +- (7,2mA + 50mA) = 360mA +- 57mA

Das Solarzellenmodul liefert einen Kurzschlussstrom von 360mA +- 57mA.

Messwertbeurteilung

In diesem Fall liegt der Messwert für eine präzise Messung ungünstig, da er im unteren Drittel des Messbereiches von 0-20A liegt. Das hat eine große Messabweichung zur Folge. Ist eine genaue Messung erforderlich, sollte ein Messgerät mit passendem Messbereich ausgewählt werden.

2.4 Spannungsmessung am Arbeitspunkt

Zur Messung der Spannung am Arbeitspunkt, wird das Solarzellenmodul mit einem Verbraucher belastet. Bei dieser Messung ist der Verbraucher ein kleiner Gleichstrommotor. Der Stromkreis ist geschlossen und es fließt ein Strom durch den Motor. Das Multimeter ist parallel zum Motor angeschlossen. An dem Innenwiderstand des Solarzellenmoduls fällt eine Spannung ab und es wird sich eine geringere Spannung als im Leerlaufbetrieb einstellen.

Das Multimeter wird auf 20V Gleichspannungsmessung eingestellt.

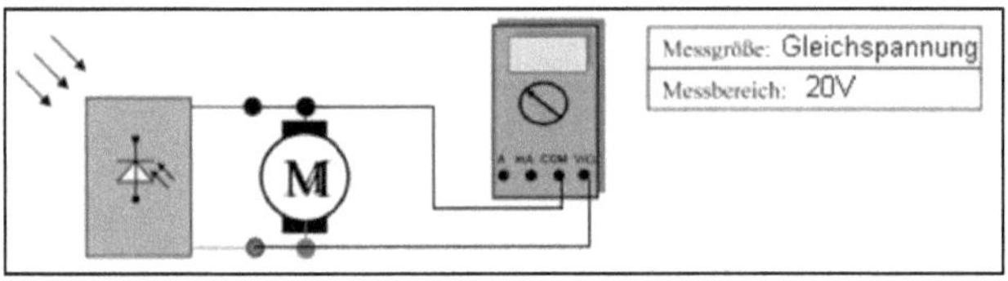

Abbildung 4: Messaufbau zur Spannungsmessung am Arbeitspunkt

Messergebnis Berechnung

$U_A = U_{Messung}$ +-(0,5% + 3d) = 2,33V +- (0,12V + 0,03V) = 2,33V +- 0,042V

Die Spannung am Betriebspunkt ist 2,33V +- 0,042V.

Messwertbeurteilung

Wie erwartet ist die Spannung im Vergleich zur Leerlaufspannung geringer. Der Spannungsabfall am Innenwiderstand des Solarzellenmoduls berechnet sich aus der Differenz von $U_{Leerlauf}$ - U_A und beträgt somit 0,09V.

2.5 Messung der abgegebenen Leistung

Es wird die abgegebene Leistung eines Solarzellenmoduls bei Belastung mit einem Gleichstrommotor gemessen werden.

Um die Spannung und den Strom gleichzeitig zu messen, werden zwei Multimeter verwendet. Die Strommessung erfolgt in Reihe zwischen dem Solarzellenmodul und dem Gleichstrommotor. Auf welcher Seite im Stromkreis das Multimeter angeschlossen wird, hat keine Auswirkung auf den Messwert. Lediglich auf die Polung muss geachtet werden, um einen positiven Messwert angezeigt zu bekommen. Die Spannungsmessung erfolgt gleich dem vorherigen Messaufbau.

Das Multimeter, welches als Strommesser angeschlossen ist, besitzt einen Innenwiderstand an dem eine geringe Spannung abfällt. Daher ist eine geringere Spannung als bei der Messung im Arbeitspunkt zu erwarten.

Der Strom ist jetzt geringer als der Kurzschlussstrom, daher wird der 200mA Messbereich eingestellt.

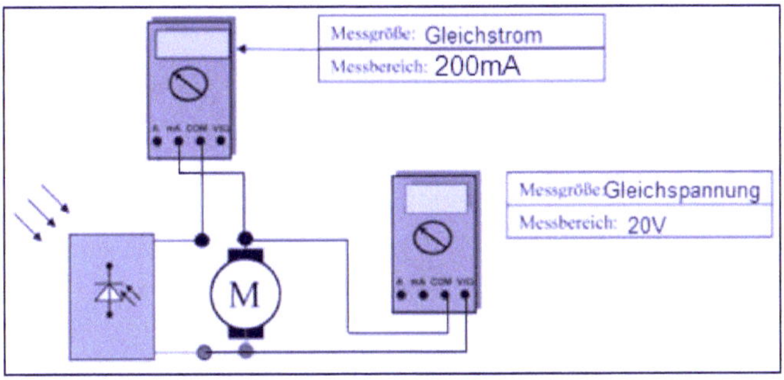

Abbildung 5: Messaufbau der Leistungsmessung

Messergebnis Berechnung

$U_A = U_{Messung}$ +-(0,5% + 3d) = 2,15V +- (0,01V + 0,03V) = 2,15V +- 0,04V

$I_A = I_{Messung}$ +-(0,8% + 3d) = 10,05mA +- (0,08mA + 0,03mA) = 10,05mA +- 0,11mA

Das Solarzellenmodul liefert am Arbeitspunkt eine Spannung von 2,15V +- 0,04V und einen Strom von 10,05mA +- 0,11mA. Damit lässt sich die abgegebene Leistung wie folgt berechnen.

$P_A = U_A * I_A = 2{,}15V * 0{,}11mA = 21{,}6mW$

Das Solarzellenmodul gibt am Arbeitspunkt eine Leistung von 21,6mW ab.

2.6 Widerstandsmessung

Es soll der Widerstand der Motorwicklung gemessen werden.

Der Motorwicklungswiderstand wird im Stillstand gemessen. Das Multimeter wird mit dem Motor verbunden und ist auf 200Ohm Widerstandsmessung eingestellt.

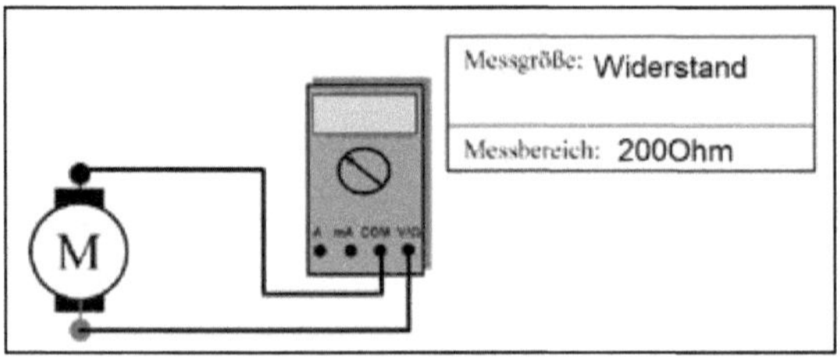

Abbildung 6: Messaufbau der Widerstandsmessung

Messergebnis Berechnung

$R_{Wicklung} = R_{Messung}$ +-(0,8% + 5d) = 45Ω +- (0,36Ω + 0,5Ω) = 45Ω +- 0,9Ω

3 Messen mit digitalem Speicher-Oszilloskop

3.1 Digital-Speicher-Oszilloskop Meteix 6062C

Mit einem digitalen Speicher-Oszilloskop kann der zeitliche Verlauf einer Spannung optisch auf einem Display dargestellt werden. Die Messung erfolgt dabei über einen A/D-Wandler. Zwei Messeingänge befinden sich an der Oberseite und sind als BNC-Buchse ausgeführt. Es ist zu beachten, dass die Masseanschlüsse elektrisch verbunden sind. Für die Messung wird das Tischoszilloskop Meteix 6062C eingesetzt.

Abbildung wurde aus datenschutzrechtlichen Gründen von der Redaktion entfernt.

Abbildung 7: Digital-Oszilloskop Meteix 6062C

3.1.1 Trigger und Single Shot

Über die Trigger Einstellung lässt sich ein stehendes Bild eines periodischen Spannungsverlaufs erzeugen. Wird der Trigger Schwellenwert überschritten startet die Datenerfassung. Liegt der Trigger Schwellwert auf einem Wert an der oberen Grenze der Amplitude, flackert das Bild. In diesem Fall ist der Trigger auf einen geringeren Wert einzustellen.

Wird die Funktion „Single Shot“ aktiviert, stoppt die Aufzeichnung nach dem Erreichen des Trigger-Schwellenwertes. Dadurch können auch einmalige Ereignisse im Spannungsverlauf betrachtet werden.

3.1.2 AC/DC/GND

Der Schalter mit der Beschriftung AC/DC/GND hat folgende Auswirkungen:

- AC - Nur die Wechselspannungsanteile der Messspannung werden dargestellt
- DC - Gleich- und Wechselspannungsanteile werden dargestellt

- GND - Messeingang wird auf GND Potential gelegt (Nützlich zum Einstellen des Nullpunkts der Vertikalablenkung)

3.2 Funktionsgenerator

Der Funktionsgenerator erzeugt Spannungssignale und bietet die Möglichkeit verschiedene Kurvenformen einzustellen. Dabei kann die Frequenz und die Amplitude variabel eingestellt werden.

Es wird der Funktionsgenerator Typ FG-7202 von Voltcraft verwendet.

3.3 Messung einer sinusförmigen Spannung

Der Messkanal 1 am Oszilloskop wird mit einem BNC-Kabel mit dem „Output“ des Funktionsgenerators verbunden.

Abbildung wurde aus datenschutzrechtlichen Gründen von der Redaktion entfernt.

Abbildung 8: Messaufbau Oszilloskop -Funktionsgenerator

Einstellungen am Funktionsgenerator

Frequenz: 50Hz Sinussignal, Amplitude: 2V

Einstellungen am Oszilloskop: vertikal Bereich 1V/div, horizontal Bereich 10ms/div

Wird die Frequenz am Funktionsgenerator erhöht, können auf dem Display mehrere Periodendurchgänge abgelesen werden. Wird sie verringert, erkennt man nur noch einen Teil der Periode. Um die Periodenzeit möglichst präzise ablesen zu können, wird daher der Horizontalbereich so eingestellt, dass genau eine Periode der Sinuswelle den Bildschirm ausfüllt.

Die 50Hz Sinuswelle hat eine Periodendauer von 20ms. Bei den vorgenommenen Einstellungen zeigt das Oszilloskop eine Periode der Sinuswelle über zwei Felder in der x-Richtung und zwei Felder in der y-Richtung an.

3.4 Messung einer rechteckförmigen Spannung

Es werden die folgenden Einstellungen am Funktionsgenerator vorgenommen:

Frequenz: 1MHz Rechtecksignal, Amplitude: 2V

Am Oszilloskop den Zeitbereich auf 1µs/div und die Vertikal Auslenkung auf 1V/div einstellen. Das Rechtecksignal wird somit auf dem Display sichtbar.

Es ist deutlich zu erkennen, dass die Flanken im Anstieg sowie beim Abfallen abgeflacht angezeigt werden. Je höher die Frequenz eingestellt wird, desto stärker verändert sich die Rechteckspannung an den Flanken.

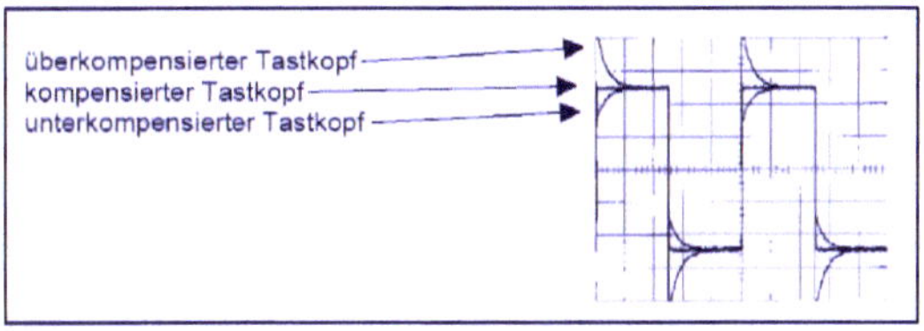

Abbildung 9: Messung einer rechteckförmigen Spannung

Aufgrund von Kapazitäten im Messaufbau muss daher eine Kompensation stattfinden. Dazu wird ein Tastkopf an das Oszilloskop angeschlossen und die automatische Kompensation durchgeführt. Alternativ kann mit der Einstellschraube am Tastkopf justiert werden.

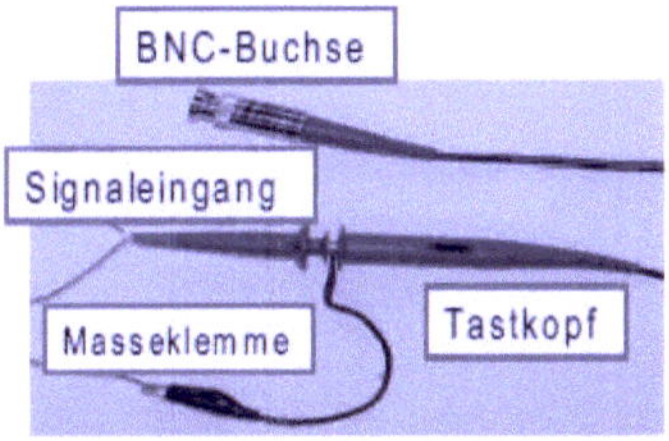

Abbildung 10: Oszilloskop Tastkopf

3.5 Spannungsmessung eines Solarzellenmoduls

Der Messkanal 1 vom Oszilloskop wird mit dem Solarzellenmodul verbunden, welches offene Kabelenden hat. Zur Verbindung wird dazu ein Tastkopf verwendet.

Einstellungen am Oszilloskop: Kopplung AC

Die Empfindlichkeit wird solange angepasst bis ein Signal zu erkennen ist.

Ergebnis: Da das Oszilloskop auf AC-Kopplung eingestellt ist, zeigt es nur den Wechselspannungsanteil des Solarzellenmoduls an. Es ist daher nur ein sehr kleines Signal messbar, welches einem Störrauschen entspricht.

4 Programmierung mit LabVIEW

4.1 LabVIEW

LabVIEW ist eine grafische Programmierumgebung der Firma National Instruments. Die Funktionalität und das Aussehen sind an reale Instrumente angelehnt. LabVIEW-Programme werden als virtuelle Instrumente oder einfach „Vi" bezeichnet. Sie bestehen aus zwei Komponenten:

Frontpanel

Das Frontpanel ist die grafische Benutzeroberfläche (GUI). Per „Drag and Drop" können die Ein- und Ausgabeelemente platziert werden.

Blockdiagramm

Das Blockdiagramm ist direkt mit den Elementen des Frontpanels verknüpft und enthält den Programmcode.

Abbildung wurde aus datenschutzrechtlichen Gründen von der Redaktion entfernt.

Abbildung 11: LabVIEW Frontpanel und dazugehöriges Blockschaltbild

Programmiert wird in LabVIEW nach dem Prinzip des Datenflusses. Sie bietet die volle Flexibilität einer modernen Hochsprache, ohne dass sich der Entwickler mit aufwendiger Syntax befassen muss. Der Programmcode im Blockdiagramm besteht aus einer Auswahl verschiedenster Elemente.

Bestandteile des Programmcodes können sein:

- Ein- und Ausgabeelemente (Controls und Indikators), sowie Konstanten, Funktionen und Operatoren
- Ablaufstrukturen (For- und While-Schleife, Case-Anweisung, ...)
- Unterprogramme (SubVi)

Diese werden miteinander verbunden und bestimmen so den Datenfluss und somit die Programmfunktion.

4.2 Erstellen und Anzeigen von Zufallszahlen

Das Vi „Zufallszahl“ liefert Zufallswerte zwischen 0 und 1. Durch die Subtraktion mit 0,5 verschiebt sich der Bereich von -0,5 bis 0,5. Dieser Wert wird in die Case-Struktur getunnelt und die boolesche Vergleichsfunktion „>= 0“ übergeben. Die Case-Struktur wird von dem Ergebnis der Vergleichsoperation gesteuert.

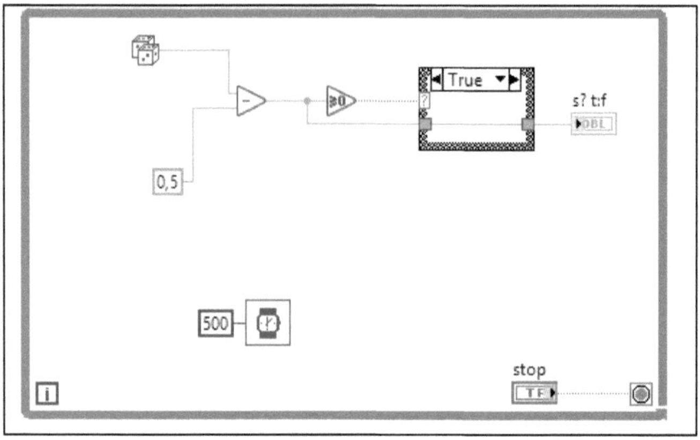

Abbildung 12: Zufallszahl - Case „“true““

Im Fall das der Wert größer oder gleich 0 ist, wird der Zufallswert ohne Veränderung durchgeleitet und auf dem Frontpanel angezeigt. Ist der Wert < 0 wird die Konstante 0 angezeigt. Nach einer Wartezeit von 500ms wird die While-Schleife neu durchlaufen, solange bis der Stopp-Button gedrückt wird.

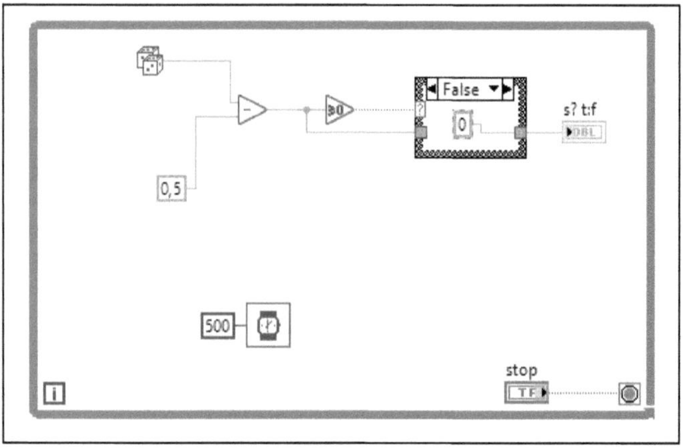

Abbildung 13: Zufallszahl - Case “false“

4.3 Inkrementieren eines Zählers mit einem Button

Zuerst werden die beiden Shift-Register mit den Konstanten "false" und 0 initialisiert.

Der OK-Button liefert im unbestätigten Zustand "false" und im betätigten Zustand "true". Solange der Button unbetätigt ist, wird das "false" Fenster der Case-Struktur ausgeführt. Der aktuell im Shift-Register hinterlegte Wert, wird dann ohne Veränderung durch die Case-Struktur geleitet, angezeigt und in das Shift-Register geschrieben. Dieser Vorgang Wiederholt sich im Takt von 1ms.

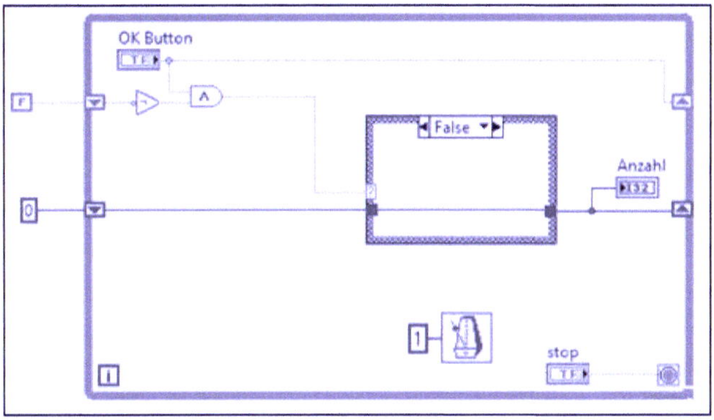

Abbildung 14: Inkrementieren eines Zählers - Case "false"

Ist der Button betätigt wird der "true" Rahmen der Case-Struktur ausgeführt. Der im blauen Shift-Register hinterlegte Wert wird um 1 inkrementiert, angezeigt und dem Shift-Register übergeben. Dadurch ist der neue Wert für den nächsten Schleifendurchlauf gespeichert.

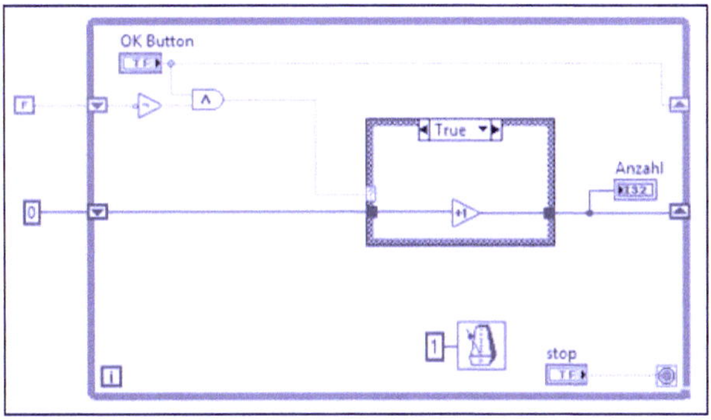

Abbildung 15: Inkrementieren eines Zählers - Case "true"

Damit der Zähler pro Button Betätigung genau um eins inkrementiert wird und es nicht aufgrund der schnellen Schleifendurchgänge zu einer mehrmaligen Inkrementierung kommt, muss eine Maßnahme getroffen werden. Dazu wurden im Labor zwei Methoden programmiert.

Methode 1: Der Button Wert wird in einem Shift-Register abgelegt. Nach erneutem Durchlaufen der While-Schleife blockiert die UND-Verknüpfung das erneute inkrementieren der Zählers solange, bis der Button wieder unbetätigt ist.

Methode 2: Dazu kann dem Button ein anderes Schaltverhalten zugewiesen werden. Die Option „Latch beim drücken" eignet sich für diesen Zweck. Wird der Button bei dieser Einstellung betätigt, setzt sich sein Wert direkt auf "false" zurück, sobald er vom Programm ausgelesen wurde.

Abbildung wurde aus datenschutzrechtlichen Gründen von der Redaktion entfernt.

Abbildung 16: Button Schaltverhalten

Es ist deutlich zu erkennen, dass die Einstellung des Schaltverhaltens weniger Programmieraufwand erfordert und dadurch auch übersichtlicher ist.

4.4 Stoppuhr

Die Stoppuhr wird über den Startbutton gestartet und über den STOP-Button angehalten. Wird erneut gestartet soll wieder bei 0 angefangen werden zu zählen.

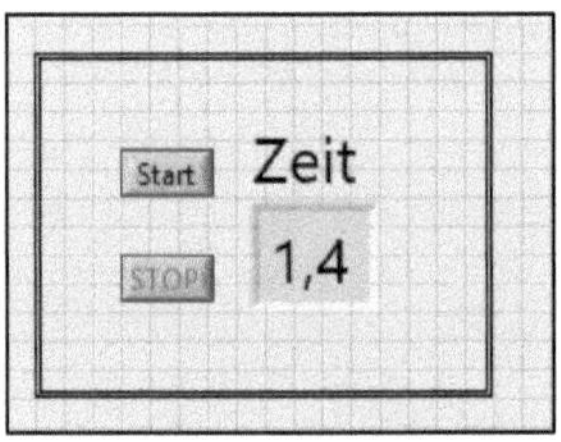

Abbildung 17: Frontpanel - Stoppuhr

Solange der Startbutton unbetätigt ist, wiederholt sich die While-Schleife im Takt von 1ms. Die Case-Struktur steht solange auf "false" und enthält keinen Programmcode. Die Wartezeit von einer Millisekunde verhindert eine hohe Auslastung des Systems.

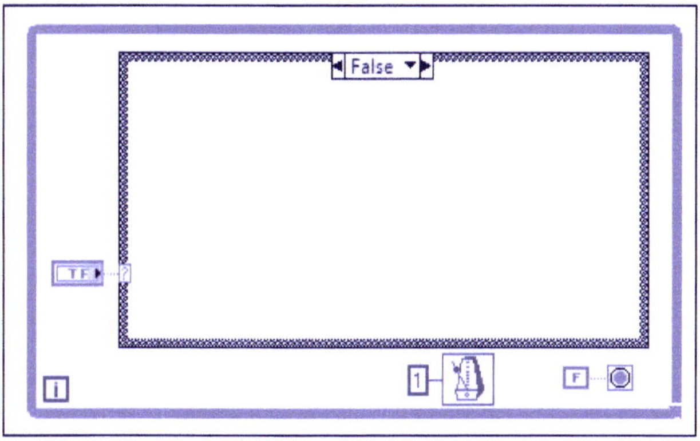

Abbildung 18: Stoppuhr - Case "false"

Ist der Startbutton betätigt, ist die Case-Struktur "true" aktiv. Das Vi "Timer-Wert (ms)" gibt die Zeit in Millisekunden aus, die seit Systemstart vergangen ist. Der Timer-Wert ist in die innere While-Schleife geführt und repräsentiert den Startwert der Zeitmessung.

In der inneren While-Schleife wird im Takt von 10ms die aktuelle Zeit abgerufen. Durch Subtraktion der aktuellen Zeit vom Startwert, ergibt sich die seit Start vergangene Zeit in Millisekunden. Im Anschluss wird der Wert durch 1000 geteilt, damit er auf dem Frontpanel in Sekunden angezeigt wird. Diese Schleife wird solange Wiederholt, bis der STOP-Button betätigt wird.

Die Wartezeit in der inneren While-Schleife definiert die Auflösung der Stoppuhr. In diesem Fall ist die Auflösung auf 0,1s festgelegt.

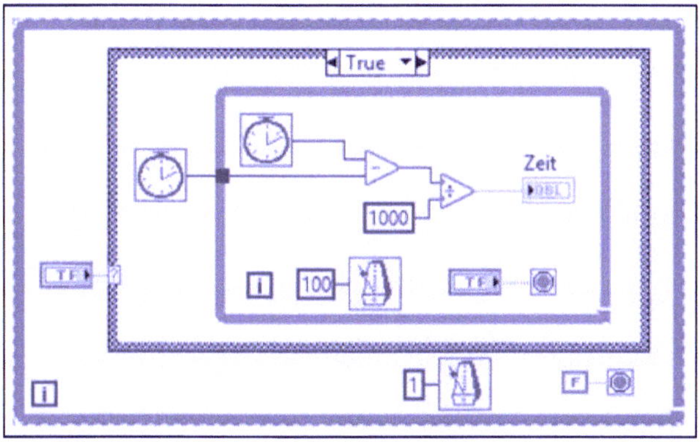

Abbildung 19: Stoppuhr - Case "true"

4.5 Würfel mit Statistik

Die Aufgabe besteht darin, einen Würfel zu erzeugen und die Anzahl der gewürfelten Augenzahlen auszuwerten.

Vorüberlegungen

Ein Würfel kann die Zahlen 1 bis 6 darstellen. Der Array Index beginnt bei 0, daher empfiehlt es sich ein Array mit 7 Elementen zu erzeugen. Dadurch kann dem Array Index die gewürfelte Augenzahl direkt zugeordnet werden. Der Array Index 0 wird nicht verwendet und bleibt immer auf dem Wert 0.

Umsetzung

Vorerst wird ein 1D-Array mit 7 Elementen erstellt und dessen Werte mit 0 initialisiert. Dieses Array beinhaltet die Anzahl der gewürfelten Augenzahlen. Es wird in die While-Schleife getunnelt.

Innerhalb der While-Schleife wird eine Zufallszahl zwischen 1 und 6 erstellt. Diese wird mit dem Vi „Zufallszahl" generiert. Dieses Vi liefert Werte zwischen 0 und 1 und muss daher auf den Bereich 1-6 angepasst werden. Daher wird mit 6 multipliziert und anschließend aufgerundet. Die Augenzahlen werden auf dem Frontpanel angezeigt.

Das Array aus dem Shift-Register wird über den Index/Augenzahl ausgelesen, um eins inkrementiert und auf denselben Index überschrieben. Die Daten in dem Array werden im Shift-Register gespeichert und zur Anzeige auf dem Frontpanel an den Waveform Graph übergeben.

Für die Statistik sind viele Werte von Vorteil, demzufolge ist die Schleifendurchlaufzeit auf 0 gesetzt.

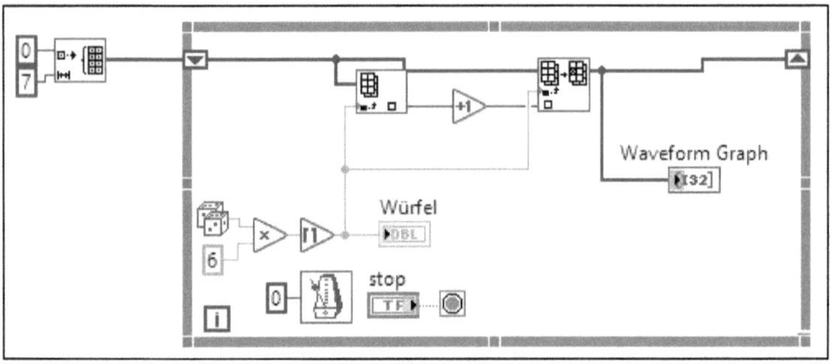

Abbildung 20: Blockdiagramm - Würfel mit Statistik

Das Balkendiagramm stellt die Verteilung der gewürfelten Zahlen dar. Es ist ersichtlich, dass jede Augenzahl gleich oft gewürfelt wurde. Das Vi „Zufallszahl“ gibt demnach keinen wirklich zufälligen Wert aus.

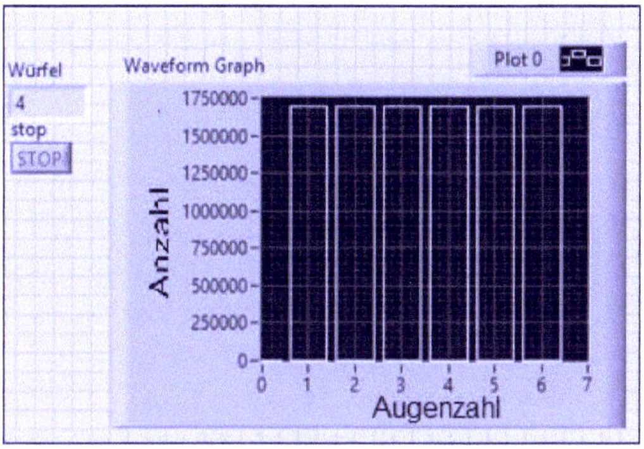

Abbildung 21: Frontpanel – Würfel mit Statistik

4.6 Umrechnung von Dezimalzahlen in das neuner Zahlensystem

Optional: Ohne Vi „Quotient & Remainder“

Auf dem Frontpanel kann eine Dezimalzahl und die Basis eines Zahlensystems eingegeben werden. Wird das Vi ausgeführt steht das Ergebnis in dem 10-stelligen Feld und das Programm stoppt. Soll ein neuer Wert berechnet werden muss das Vi wieder neu gestartet werden.

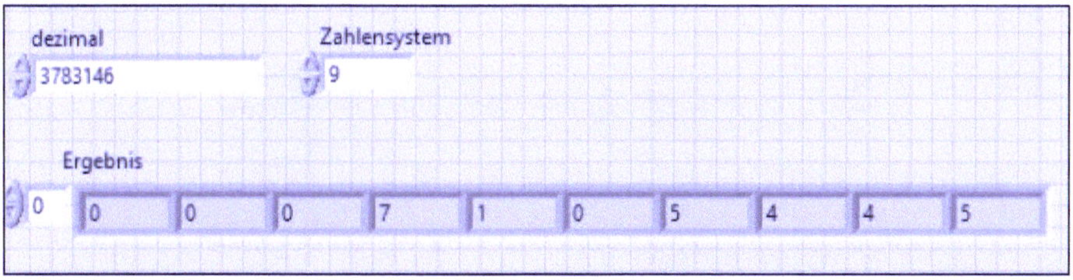

Abbildung 22: Frontpanel - Umrechnung von Dezimalzahlen

4.6.1 Mathematische Umrechnung

Die Dezimalzahl wird mit der Basis des Zahlensystems dividiert. Der Rest der Division ergibt die niederwertigste Stelle der Zahl im neuen Zahlensystem. Der Ganzzahlquotient der Division wird wieder mit der Basis des Zahlensystems dividiert und dessen Rest ergibt die nächst höherer Stelle der Zahl im neuen Zahlensystem.

Diese Berechnung wird so oft wiederholt, bis das Ergebnis der Division 0 ergibt oder je nach Anzahl der gewünschten Stellen der Umrechnung.

4.6.2 Umsetzung in LabVIEW

Zuerst wird die im Frontpanel eingegebene Dezimalzahl in die For-Schleife geführt. Der Wert aus dem Shift-Register wird mit der Basis des im Frontpanel eingetragenen Zahlensystems dividiert. Um ein ganzzahliges Ergebnis zu erhalten, wird das Ergebnis abgerundet. Mit diesem Wert wird im neuen Schleifendurchgang weiter gerechnet, daher wird er in das Shift-Register abgelegt.

Der ganzzahlige Wert multipliziert mit der Basis des gewählten Zahlensystems, wird von dem Wert aus dem Shift-Register abgezogen. Das Ergebnis ist der Rest, welcher über ein Array aus der For-Schleife herausgeführt wird. LabVIEW legt dieses Array selbstständig an.

Die For-Schleife wird zehnmal durchgeführt, um zehn Stellen des Ergebnisses zu berechnen. Damit die niederwertigste Stelle ganz rechts im Array angezeigt wird, dreht das Vi „Reverse 1D Array" die Reihenfolge um.

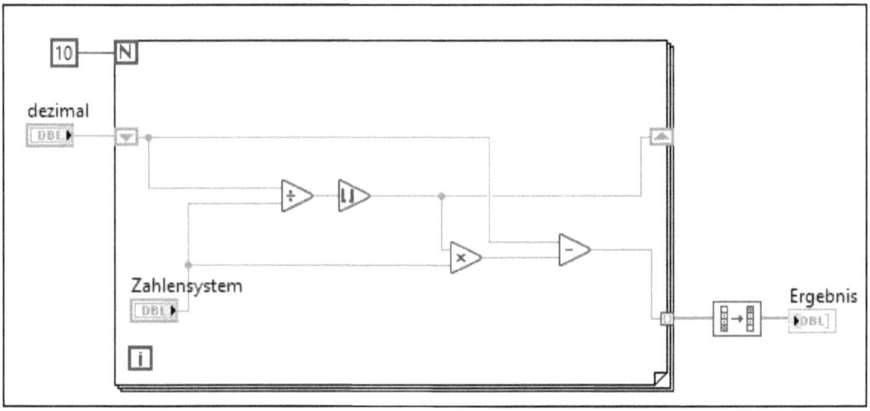

Abbildung 23: Blockdiagramm - Umrechnung von Dezimalzahlen

5 Messen und Auswerten mit LabVIEW

5.1 Datenerfassungskarte USB-6008

Zur Messwerterfassung mit LabVIEW steht im Labor eine Multifunktions-Datenerfassungskarte USB-6008 zur Verfügung. Diese verfügt über diverse analoge Eingänge und digitale Ein-/Ausgänge.

Abbildung 24: Multifunktionskarte USB-6008

5.1.1 Abtastrate

Die maximale Abtastrate der Multifunktionskarte ist mit 10kS/s im Datenblatt angegeben. Nach dem Gesetz von Shannon-Hartley, darf die Frequenz des Messsignals daher 5khz nicht überschreiten.

5.2 Kalibrierung eines Wegsensors inkl. Parameter

5.2.1 Messaufbau

Die Messübertragung erfolgt im Single-Ended-Mode, das Messsignal liegt dabei auf einer Anschlussseite auf Masse. Das Spannungssignal $U_{Messung}$ wird an Terminal 2 der Datenerfassungskarte angeschlossen, GND an Terminal 1, Spannungsversorgung an Terminal 31.

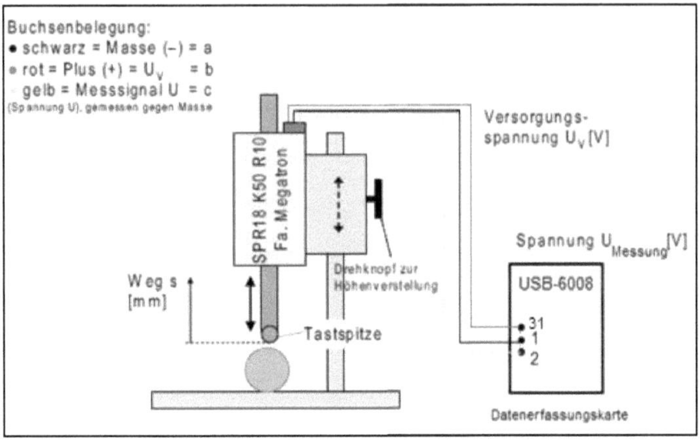

Abbildung 25: Messaufbau einer Wegmessung

5.2.2 Aufnahme der Sensorkennlinie

Im „Measurement & Automation Explorer“ wird ein neuer Task angelegt. Nach Einstellung des Messbereichs von 0 bis 5V, kann die Messspannung direkt abgelesen werden.

Es werden nacheinander kalibrierte 10mm hohe Endmaße zwischen Tastspitze und Messauflage gelegt. Für den Bereich 0-40mm werden die Spannungswerte in die Tabelle eingetragen.

Tabelle 1: Wegsensor Kalibrierwerte

Anzahl Endmaße	Weg s [mm] von Hand eingetragen	Spannung U [V] mit USB-6008 gemessene
0	0	0,01576
1	10	1,03472
2	20	2,04348
3	30	3,04206
4	40	4,0712
(5)	(50)	—

5.2.3 Umsetzung in LabVIEW

Mittels DAQ-Assistenten kann der analoge Eingang in LabVIEW eingefügt werden. Dort wurde folgendes eingestellt: Messbereich = 0-5V, Messart ="RSE"(Single-Ended), Acquisition Mode = „Aquire Continously“, Samples to read = 100 und die Abtastrate = 1000Hz

Das bedeutet, dass die Datenerfassungskarte die physikalische Messgröße 1000-mal in einer Sekunde abtastet und diese im Pufferspeicher ablegt. Sobald 100 Werte im Pufferspeicher liegen, werden sie als Datenpaket an LabVIEW weitergeleitet. LabVIEW erhält demzufolge im Takt von 100ms jeweils 100 Messwerte.

Die Parameter für den Weg in mm und die dazugehörigen Spannungswerte in V, können in die beiden Arrays eingetragen werden. Dafür werden zunächst die Arrays mit „Array Constant“ erstellt und mit jeweils fünf Konstanten gefüllt.

Mit der Funktion „Append Array“ wird ein 2D Array gebildet. Es werden zwei 2D Arrays erzeugt, eines davon erhält die linearisierten Spannungswerte aus dem Vi „Linear Fit“. Diese 2D Arrays werden zu einem Cluster zusammengefügt und dem Graph übergeben. Folglich zeigt der Graph beide Kennlinien an.

Der Weg *s* berechnet sich nach folgender Formel: $s = \frac{U_{Messung} - U_0}{k}$

Das Vi „Linear Fit“ stellt die Werte Verstärkungsfaktor *k* und Offset U_0 zur Verfügung. Die Spannung $U_{Messung}$ wird vom Vi „DAQ Assistent“ geliefert. Mit diesen Werten kann die Berechnung von Weg *s* durchgeführt werden.

Abbildung wurde aus datenschutzrechtlichen Gründen von der Redaktion entfernt.

Abbildung 26: Blockdiagramm - Kalibrierung eines Wegsensors

Auf dem Frontpanel sind die Messwerte sichtbar. In dem Spannungs-Weg-Graph sind die linearisierte Kennlinie (grau) sowie die Kalibrierwerte (rot) zu sehen. Die Kalibrierwerte können zur besseren Verdeutlichung verändert werden.

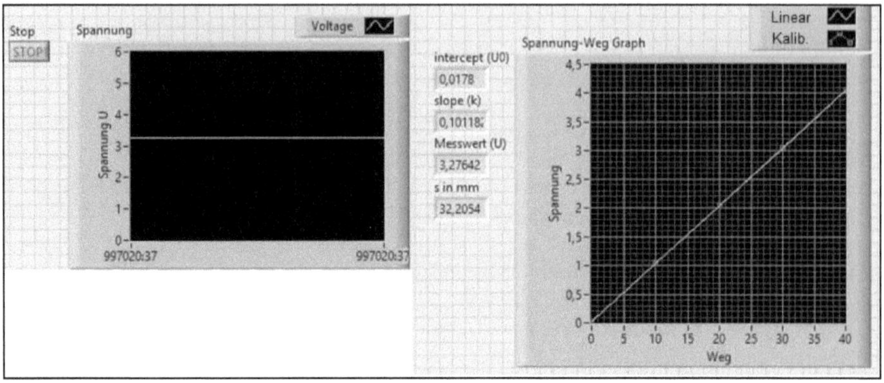

Abbildung 27: Frontpanel - Kalibrierung eines Wegsensors

5.3 Kalibrierung eines Temperatursensors (Unterschiede zum Wegsensor)

Der Temperatursensor wird an die Datenerfassungskarte angeschlossen.

Tabelle 2: Anschluss Temperatursensor - Datenerfassungskarte

Temperatursensor	Datenerfassungskarte
Schwarz GND	Terminal 1 GND
Weiß Output	Terminal 2 AI0
Rot V_{DD}	Terminal 31 +5V

Der Temperatursensor gibt im Messbereich zwischen -40 bis 125°C bereits ein lineares Signal aus, die Arrays und die lineare Regression werden deshalb entfernt. Stattdessen werden Konstanten für den Offset- und die Verstärkung angelegt. In dem Datenblatt sind diese Werte zu finden.

Abbildung wurde aus datenschutzrechtlichen Gründen von der Redaktion entfernt.

Abbildung 28: Blockdiagramm - Kalibrierung eines Temperatursensors

5.4 Auswertung einer Lichtschranke

Im Labor wurde eine Gabellichtschranke bereitgestellt, diese war bereits mit drei Adern verlötet und konnte daher direkt an die Terminals angeschlossen werden. Ist die Lichtschranke unterbrochen, liegt ein „High"-Signal am Signalausgang an, andernfalls liegt das Signal auf GND.

Tabelle 3: Anschluss Lichtschranke - Messdatenerfassungskarte

Lichtschranke Aderfarbe	Datenerfassungskarte
Schwarz - GND	Terminal 1 - GND
Gelb – Signal-Output	Terminal 17 - P0.0
Rot - V_{DD}	Terminal 31 - +5V

Zuerst wird die Datenerfassungskarte initialisiert, indem „Dev1/port0/line0" ausgewählt und der Kanal P0.0 als digitaler Eingang definiert wird.

In der While-Schleife wird der Wert des Eingangs ständig ausgelesen und der LED übergeben. Über den STOP-Button wird die While-Schleife verlassen und die Verbindung zur Datenerfassungskarte wieder freigeben.

Abbildung wurde aus datenschutzrechtlichen Gründen von der Redaktion entfernt.

Abbildung 29: Blockdiagramm - Auswertung einer Lichtschranke

Ist die Lichtschranke unterbrochen leuchtet die LED auf dem Frontpanel.

Abbildung wurde aus datenschutzrechtlichen Gründen von der Redaktion entfernt.

Abbildung 30: Frontpanel - Auswertung einer Lichtschranke